Cover photo: Drawings of a fruit and seeds studied.

This book reflects the results contained in my Bachelor's thesis in Biology at the Central University of Venezuela (UCV) in Caracas in 1980: "Morphological and anatomical study of fruits and seeds of Venezuelan species of the Asclepiadaceae family" (in Spanish).

The former Asclepiadaceae family is currently the Asclepiadoideae subfamily of the Apocynaceae family.

All species studied here was vines, with the exception of the erect herb *Asclepias curassavica*. Their global morphology is the same as that of the South American vine *Araujia sericifera* Brot. (Apocynaceae), currently naturalized in warmer parts of Europe (figures 1-6, 16-66).

The fruit is composed of 1-2 dehiscent follicles. The follicle has a ventral placenta with a zone of laminar longitudinal outgrowths.

The seeds are ovate in shape and dorso-ventrally compressed. They have an inconspicuous to very wide aliform projection. They have a white and silky tuft of unicellular and cellulosic hairs (except *Marsdenia rubrofusca* and *Matelea delascioi*). They are brown to black in color.

The endosperm is cartilaginous, whitish, totally enveloping the embryo.

The embryo is differentiated, straight, axial. The cotyledons are thin, non-photosynthetic, full-edged and with apparent veining. The radicle is conical, not covered by cotyledons.

The anatomical cross-sections of the seeds (Figures 7-15) show a basic organization: a unistrata epidermis of cells covered by a cuticle and containing tannin, and a second layer of parenchymal cells with strongly thickened (non-lignified) and obliterated walls, and with rhombohedral crystals (observed in most of the species studied). In the aliform projection, the parenchymal cells may or may not be obliterated.

The endosperm is made up of fat-content parenchymal cells. The embryo is made up of a unistrate upper epidermis, a unistrate lower epidermis, with cells smaller than the cells of the upper epidermis, a mesophyll made up of a unistrate palisade parenchyma and a spongy parenchyma. The procambium cords and abundant fat as a reserve are also observed.

The dispersion of seeds is certainly essentially aerial (with the exception of *Marsdenia rubrofusca* and *Matelea delascioi*):

A dehiscent follicle, which gradually opens due to the drying of the air; presence in the seeds of an apical tuft of hairs; dorso-ventral compression of the seeds. This compression (according to Van der Pijl, L.. 1972. Principles of Dispersal In Higher

Plants. Springer-Verlag), gives the seed a large surface area and a relatively small weight. When the compression is less, the seed has the shape of a "drop of water", with the dorsal surface convex and the ventral face with numerous outgrowths, generally papillate. The weight of the seed is relatively small and the presence of numerous cells with reticulated thickenings in the outgrowths of the seed coat probably decreases the ratio of weight per unit volume.

These morphological characteristics of the seeds coincide with the environment in which most of the species studied live, an environment in which there is, in general, a marked delimitation between the rainy season and the dry season, which can be relatively long lasting (other species grow in secondary vegetation in more humid environments).

In *Marsdenia rubrofusca* and *Matelea delascioi*, which do not have a tuft of hairs in their relatively voluminous and heavy seeds, and which live in an environment of heavy annual rains, periodically flooded and with a dense hydrographic network, water seems to have replaced air as the basic dispersing agent of their seeds.

Figure 1. *Araujia sericifera.* Fruit. Bayonne 1997.

Figure 2. *Araujia sericifera*. Fruit and seeds. Bayonne. 1997.

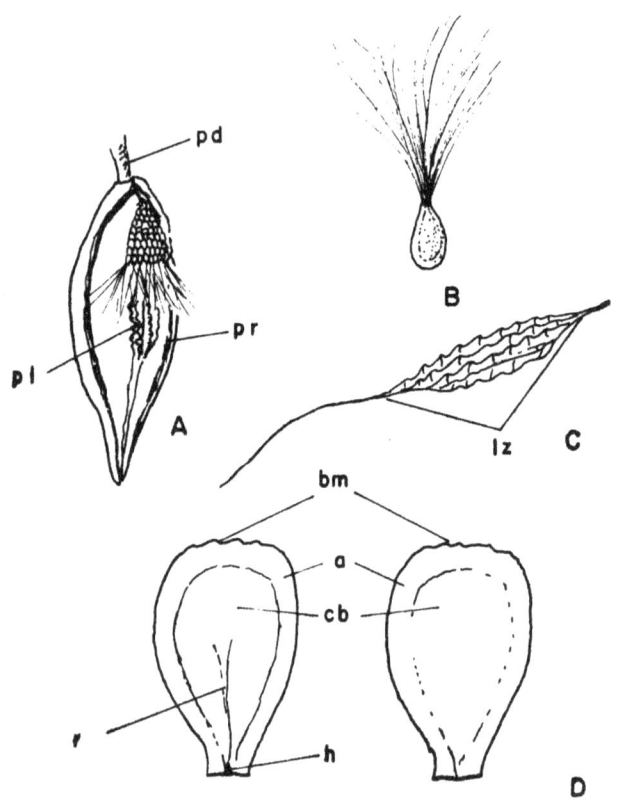

Figure 3: A. Fruit of *Blepharodon mucronatum*: pd: pedicel ; pl: placenta; pr: pericarp. B. Seed with its tuft of hairs of *Asclepias curassavica*. C. Placenta of *Blepharodon mucronatum*: zl: zone of laminar outgrowths. D. Seed of *Gonolobus rostratus* (ventral and dorsal faces): h: hilum; bm: basal margin; r: raphe; a: aliform projection; cb: central body.

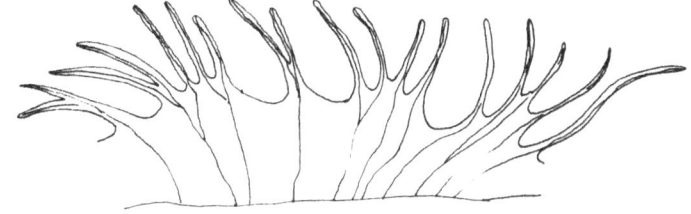

Figure 4. *Gonolobus lasiostomus*. A lamina of the placenta of the fruit.

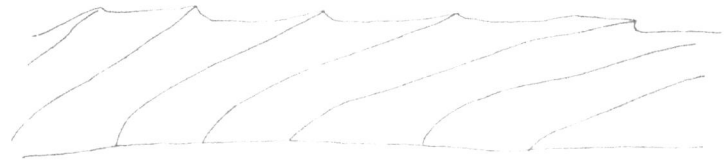

Figure 5. *Marsdenia undulata*. A lamina of the placenta of the fruit.

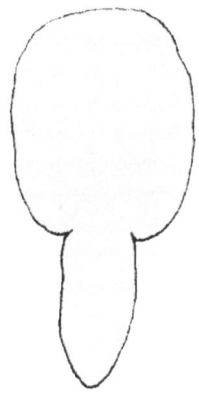

Figure 6. *Asclepias curassavica.* Embryo.

	PAGES
LIST OF SPECIES STUDIED	13
SIZES OF FRUITS AND SEEDS	15
ANATOMY OF SEED	17-22

PHOTOGRAPHS OF FRUITS AND SEEDS	23-50
Asclepias curassavica	23
Blepharodon mucronatum	24
Blepharodon nitidum	25
Cynanchum mucronatum	26-27
Cynanchum parviflorum	28
Ditassa oxyphylla	29
Gonolobus lasiostomus	30-31
Gonolobus rostratus	31-33
Marsdenia condensiflora	33-34
Marsdenia macrophylla	34-35
Marsdenia rubrofusca	35-36
Marsdenia undulata	36-37

Marsdenia xerohylica	37-38
Matelea delascioi	38-39
Matelea humboldtiana	39-40
Matelea maritima	40-41
Matelea planiflora	41-42
Matelea trianae	42-43
Matelea urceolata	43-44
Matelea viridiflora	44-45
Metalepis albiflora Urb.	46
Nephradenia linearis	47
Oxypetalum cordifolium	48
Sarcostemma clausum	49
Sarcostemma glaucum	50

List of species studied:

Asclepias curassavica L.

Blepharodon mucronatum (Schltdl.) Decne.

Blepharodon nitidum (Vell.) J.F.Macbr.

Cynanchum mucronatum (Decne.) Reiche

Cynanchum parviflorum Sw.

Ditassa oxyphylla Turcz.

Gonolobus lasiostomus Decne.

Gonolobus rostratus (Vahl) R. Br.

Marsdenia condensiflora S.F. Blake

Marsdenia macrophylla (Humb. & Bonpl. ex Schult.) E. Fourn.

Marsdenia rubrofusca E. Fourn.

Marsdenia undulata (Jacq.) Dugand

Marsdenia xerohylica Dugand

Matelea delascioi Morillo

Matelea humboldtiana Spellman & Morillo

Matelea maritima (Jacq.) Woodson

Matelea planiflora (Jacq.) Dugand

Matelea trianae (Decne. ex Triana) Spellman

Matelea urceolata (H. Karst.) L.O. Williams

Matelea viridiflora (G. Mey.) Woodson

Metalepis albiflora Urb.

Nephradenia linearis Benth. ex E. Fourn.

Oxypetalum cordifolium (Vent.) Schltr.

Sarcostemma clausum (Jacq.) Schult.

Sarcostemma glaucum Kunth

Sizes of fruits and seeds.

SPECIES	FRUIT (cm)	SEED (mm)
Asclepias curassavica	5.88 – 10.1 x 1.61	5.2 x 3.5
Blepharodon mucronatum	7.13 – 9.17 x 2.49 – 2.8	7.0 -7.6 x 2.2 -2.4
Blepharodon nitidum	7.09 – 9.9 x 1.53 - 1.69	5.2 – 6.1 x 2.9 – 3.1
Cynanchum mucronatum	2.61 – 3.61 x 0.91	5.0 – 6.3 x 2.4 – 2.8
Cynanchum parviflorum	4.5 – 8.41 x 0.12 – 0.72	4.7 – 5.1 x 1.0 – 1.2
Ditassa oxyphylla	4.3 -6.03 x 1.09 – 1.99	5.2 – 6.0 x 2.9 – 3.1
Gonolobus lasiostomus	13.5 – 17.0 x 9.3 – 11.6	6.7 – 10.1 x 3.2 – 4.5
Gonolobus rostratus	15.2 – 21.0 x 8.7 – 12.5	3.5 – 11.2 x 4.0 – 4.9
Marsdenia condensiflora	9.06 – 13.0 x 3.5 – 3.7	8.8 – 12.0 x 5.0 – 6.3
Marsdenia macrophylla	12.4 – 18.1 x 5.05 – 5.49	11.3 – 12.5 x 7.0 – 7.8
Marsdenia rubrofusca	9.3 – 12.4 x 4.6 – 5.1	34.1 – 37.0 x 18.9 – 22.6

Species		
Marsdenia undulata	12.5 – 16.0 x 5.1 – 5.3	11.4 – 13.5 x 8.2 – 9.2
Marsdenia xerohylica	10.2 – 11.1 x 4.35 – 4.72	8.4 – 10.2 x 5.8 -6.2
Matelea delascioi	10.6 x 3.0	20.5 – 23.1 x 11.3 – 11.8
Matelea humboldtiana	8.45 – 13.8 x 4.88 – 8.15	7.8 – 8.9 x 5.3 – 6.3
Matelea maritima	6.1 – 9.0 x 3.1 – 5.9	5.4 – 5.7 x 2.5 – 2.8
Matelea planiflora	6.54 – 8.73 x 2.25 – 2.81	6.4 – 7.8 x 2.8 – 3.2
Matelea trianae	11.0 x 5.5	9.1 -10.3 x 5.3 – 6.8
Matelea urceolata	10.9 – 13.1 x 4.95 – 5.74	7.8 – 8.5 x 5.4 – 5.8
Matelea viridiflora	8.1 – 9.0 x 3.25 – 3.52	5.8 – 6.0 x 3.2- 3.3
Metalepis albiflora	19.0 x 7.0	14.2 – 15.8 x 9.2 – 9.9
Nephradenia linearis	4.0 x 0.55	5.1 x 3.2
Oxypetallum cordifolium	7.1 x 2.54	6.8 – 7.5 x 3.0 – 3.4
Sarcostemma clausum	5.58 – 7.85 x 1.62 – 2.27	3.7 – 3.9 x 2.0
Sarcostemma glaucum	8.95 – 11.3 x 1.31 – 1.76	5.4 – 5.8 x 2.6 – 2.7

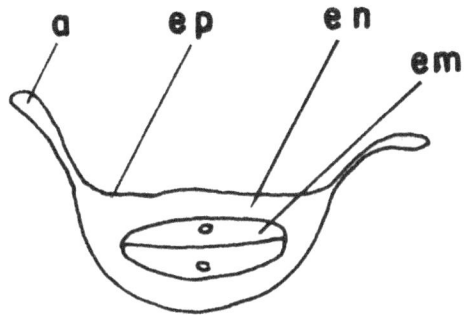

Figure 7. *Asclepias curassavica*. Seed cross-section: a: aliform projection; ep: epidermis; en: endosperm; em: embryo.

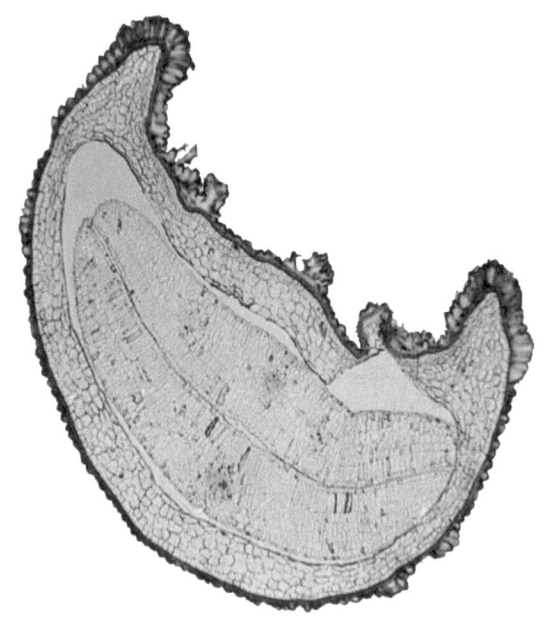

Figure 8. *Cynancum parviflorum.* Seed cross-section.

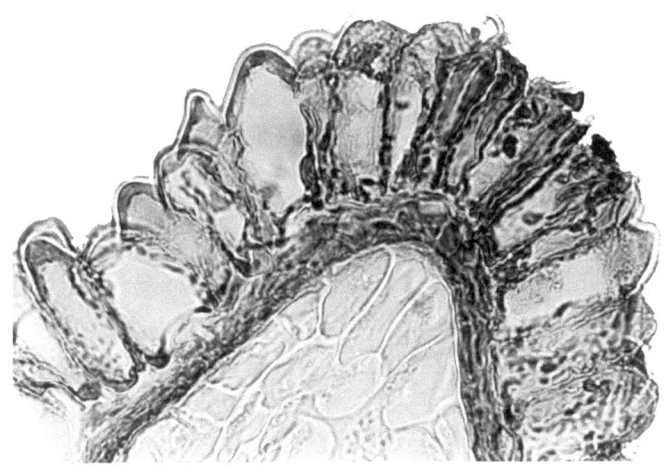

Figure 9. *Cynanchum parviflorum*. Seed cross-section, with crystals in the seed coat.

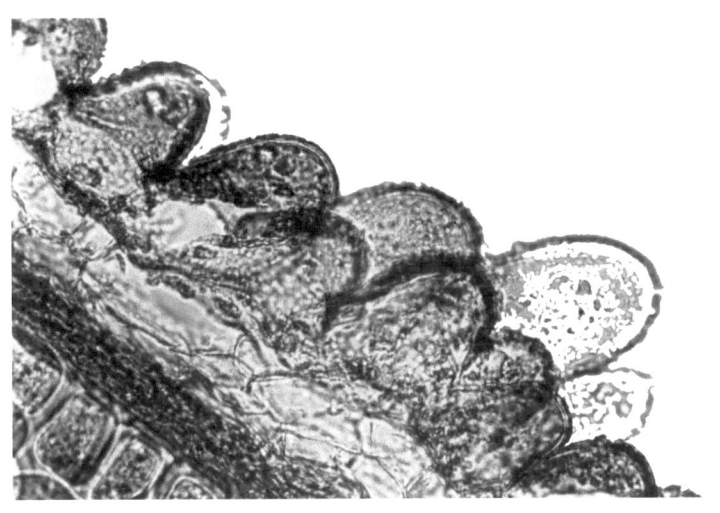

Figure 10. *Matelea planiflora*. Seed cross-section.

Figure 11. *Asclepias curassavica*. Seed cross-section.

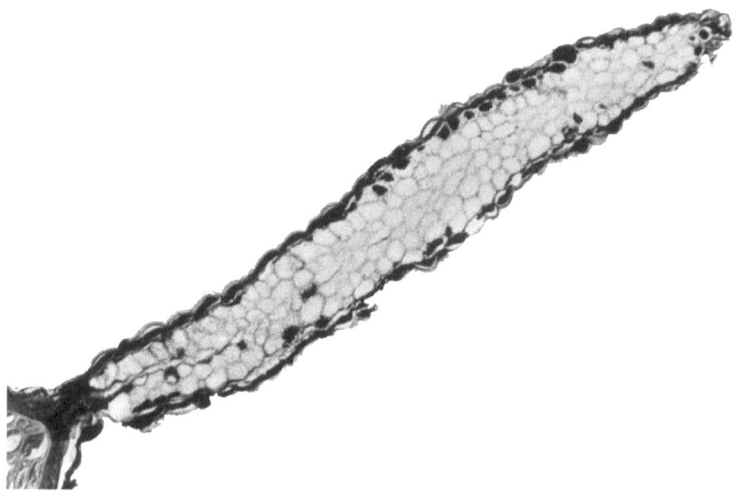

Figure 12. *Matelea urceolata*. Seed cross-section.

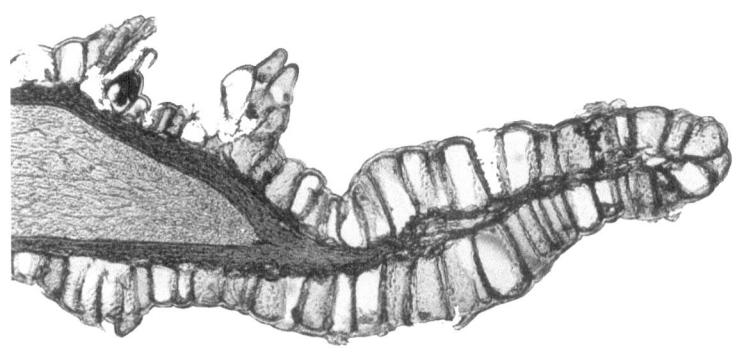

Figure 13. *Nephradenia linearis.* Seed cross-section.

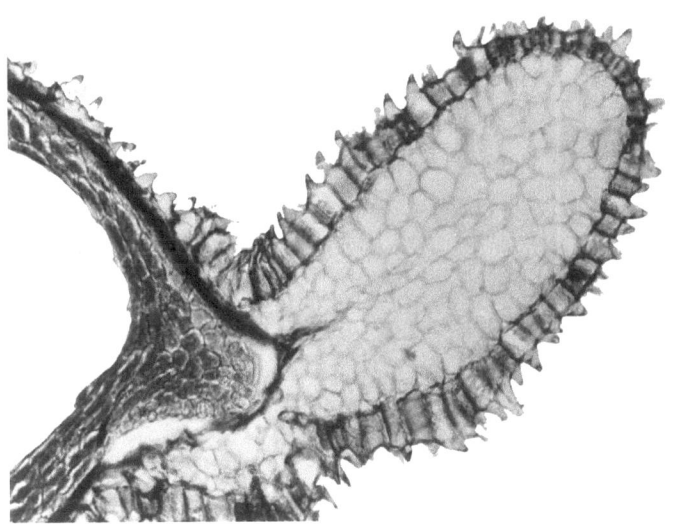

Figure 14. *Sarcostemma clausum.* Seed cross-section.

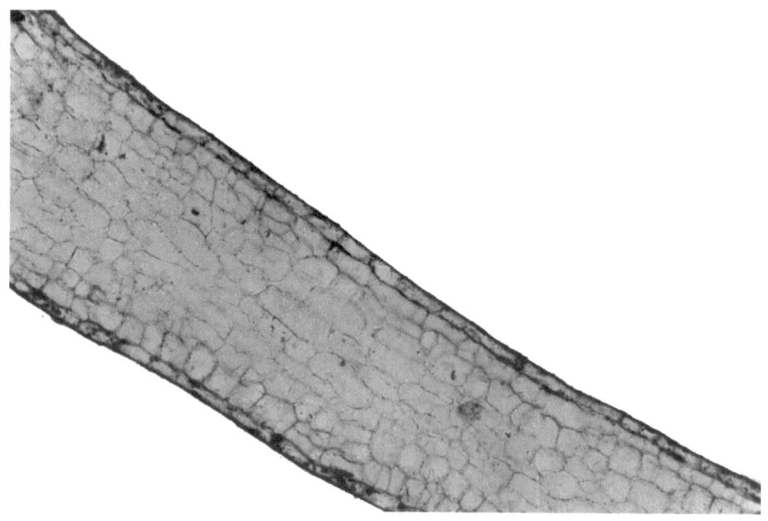

Figure 15. *Marsdenia rubrofusca*. Seed cross-section. Aliform projection.

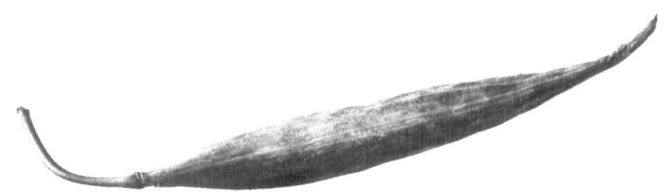

Figure 16. *Asclepias curassavica*. Fruit.

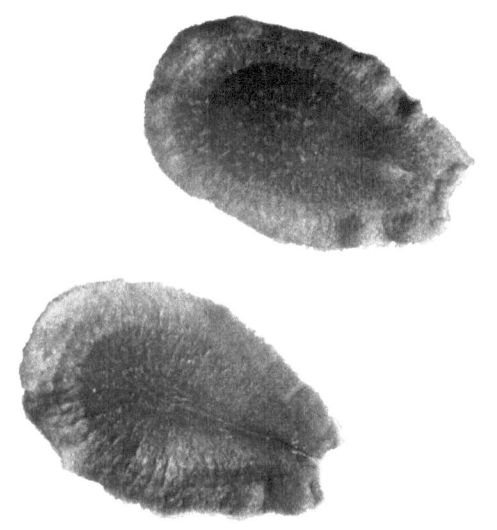

Figure 17. *Asclepias curassavica.* Seeds.

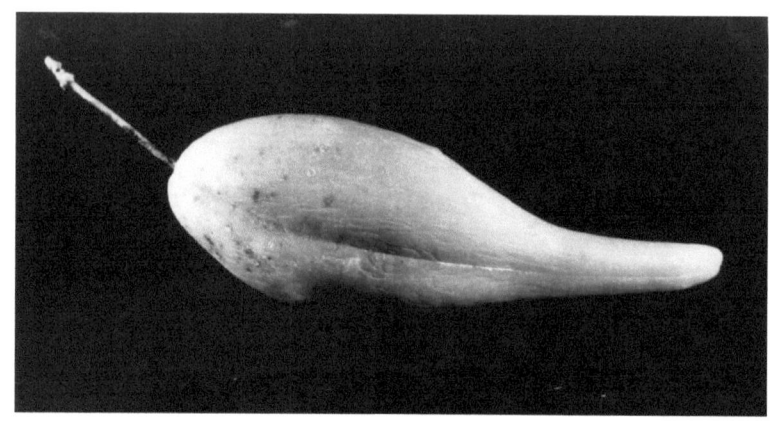

Figure 18. *Blepharodon mucronatum*. Fruit.

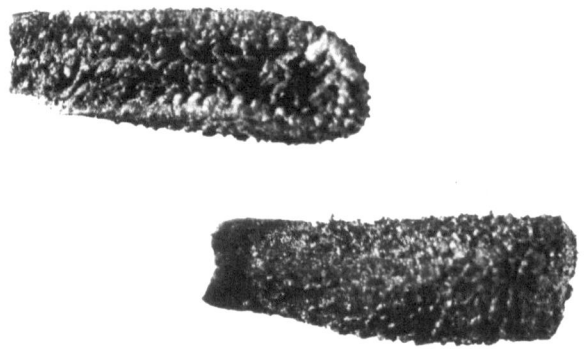

Figure 19. *Blepharodon mucronatum*. Seeds.

Figure 20. *Blepharodon nitidum*. Fruit.

Figure 21. *Blepharodon nitidum*. Seeds.

Figure 22. *Cynanchum mucronatum*. Fruit.

Figure 23. *Cynanchum mucronatum*. Seeds.

Figure 24. *Cynanchum parviflorum.* Fruit.

Figure 25. *Cynanchum parviflorum.* Seeds.

Figure 26. *Ditassa oxyphylla*. Fruit.

Figure 27. *Ditassa oxyphylla.* Seeds.

Figure 28. *Gonolobus lasiostomus*. Fruit.

Figure 29. *Gonolobus lasiostomus.* Seeds.

Figure 30. *Gonolobus rostratus.* Fruit.

Figure 31. *Gonolobus rostratus.* Fruit.

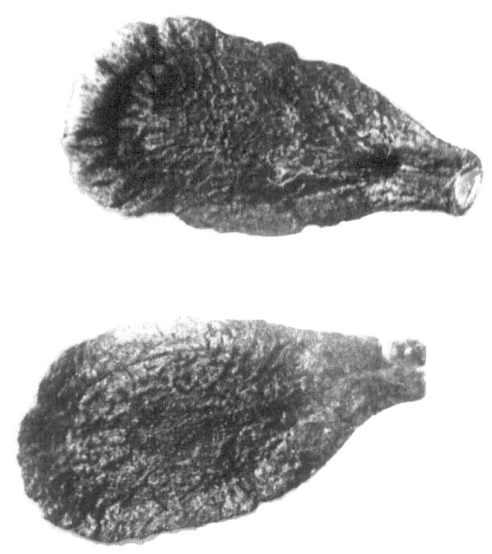

Figure 32. *Gonolobus rostratus.* Seeds.

Figure 33. *Marsdenia condensiflora.* Fruit.

Figure 34. *Marsdenia condensiflora*. Seeds.

Figure 35. *Marsdenia macrophylla*. Fruit.

Figure 36. *Marsdenia macrophylla*. Seeds.

Figure 37. *Marsdenia rubrofusca*. Fruit.

Figure 38. *Marsdenia rubrofusca*. Seeds.

Figure 39. *Marsdenia undulata*. Fruit.

Figure 40. *Marsdenia undulata.* Seeds.

Figure 41. *Marsdenia xerohylica.* Fruit.

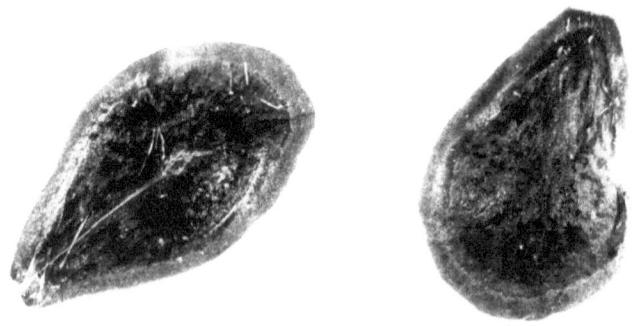

Figure 42. *Marsdenia xerohylica*. Seeds.

Figure 43. *Matelea delascioi*. Fruit.

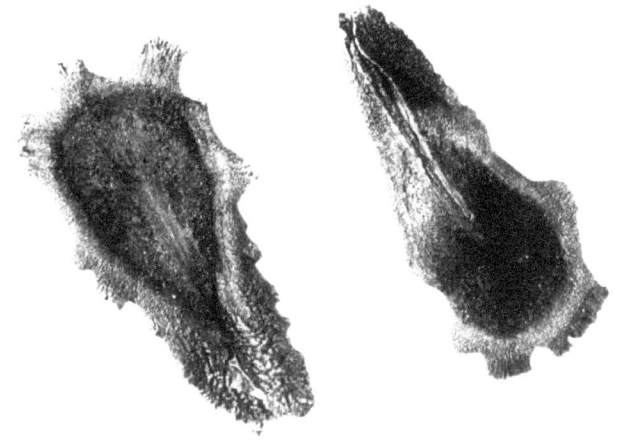

Figure 44. *Matelea delascioi*. Seeds.

Figure 45. *Matelea humboldtiana*. Fruit.

Figure 46. *Matelea humboldtiana*. Seeds.

Figure 47. *Matelea maritima*. Fruit.

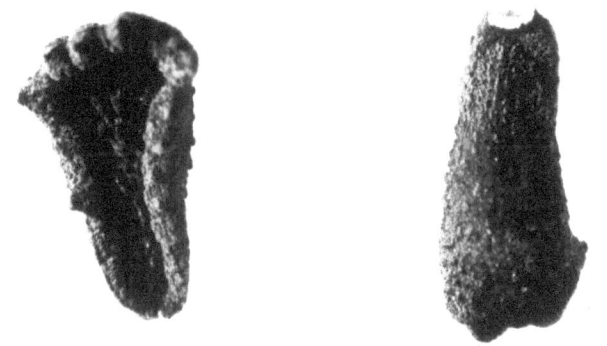

Figure 48. *Matelea maritima*. Seeds.

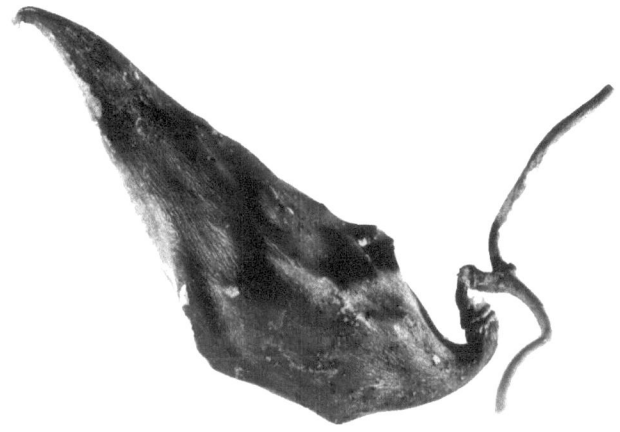

Figure 49. *Matelea planiflora*. Fruit.

Figure 50. *Matelea planiflora*. Seeds.

Figure 51. *Matelea trianae*. Fruit.

Figure 52. *Matelea trianae*. Seeds.

Figure 53. *Matelea urceolata*. Fruit.

Figure 54. *Matelea urceolata*. Seeds.

Figure 55. *Matelea viridiflora*. Fruit.

Figure 56. *Matelea viridiflora*. Seeds.

Figure 57. *Metalepis albiflora*. Fruits.

Figure 58. *Metalepis albiflora*. Seeds.

Figure 59. *Nephradenia linearis*. Fruit.

Figure 60. *Nephradenia linearis*. Seeds.

Figure 61. *Oxypetalum cordifolium*. Fruit.

Figure 62. *Oxypetalum cordifolium*. Seeds.

Figure 63. *Sarcostemma clausum*. Fruit.

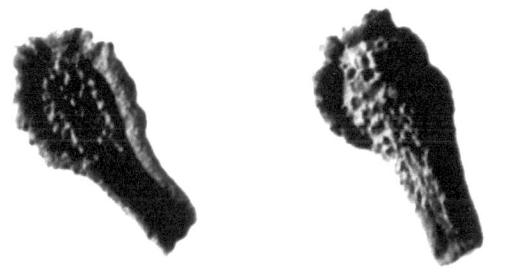

Figure 64. *Sarcostemma clausum*. Seeds.

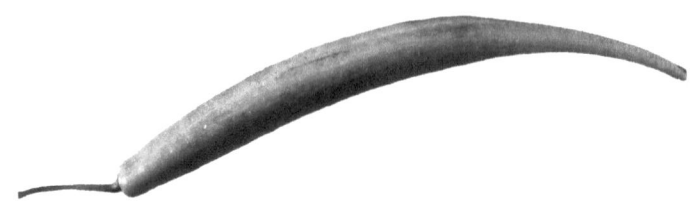

Figure 65. *Sarcostemma glaucum*. Fruit.

Figure 66. *Sarcostemma glaucum*. Seeds.

www.ingramcontent.com/pod-product-compliance
Lightning Source LLC
Chambersburg PA
CBHW040245220526

45473CB00001B/370